Improving the Army Planning, Programming, Budgeting, and Execution System (PPBES)

The Planning Phase

Leslie Lewis

Harry Thie

Roger Brown

John Schrader

Prepared for the United States Army

RAND

Arroyo Center

For more information on the RAND Arroyo Center, contact the Director of Operations, (310) 393-0411, extension 6500, or visit the Arroyo Center's Web site at http://www.rand.org/organization/ard/

PREFACE

This report documents the work of RAND's Arroyo Center on the planning phase of the Army Planning, Programming, Budgeting, and Execution System (PPBES). Arroyo Center researchers were asked to assess the effectiveness of the reengineering of the Army planning and programming process in fiscal years (FY) 1995 and 1996. The Army had modified its planning and programming documents and asked the Arroyo Center to assess several of them to determine the extent to which the reengineering was successful and to suggest improvements. The most important document is The Army Plan (TAP), the document that links planning to programming and provides the initial programming guidance to the Army Program Evaluation Groups (PEGs). It is the primary focus of this report.

This report also documents RAND's assessment and recommendations for TAP 2000–2015. This report should be of interest to those in the Department of Defense and in the Department of the Army involved in planning and programming.

The research was sponsored by the U.S. Army's Deputy Chief of Staff for Operations and Plans (DCSOPS) and was conducted in RAND Arroyo Center's Strategy, Doctrine, and Resources Program. The Arroyo Center is a federally funded research and development center sponsored by the United States Army.

CONTENTS

FIGURES

TABLES

BACKGROUND AND OBJECTIVES

During FY95 and FY96, the Army reengineered its programming process. The Army also modified its planning and programming documents, such as the Army Strategic Planning Guidance (ASPG), The Army Plan (TAP), the Army Programming Guidance Memorandum (APGM), and the Program Objectives Memorandum (POM) to reflect the new process. In 1997, the Army asked RAND Arroyo Center to assess TAP for 2000–2015 to determine the extent to which reengineering efforts were reflected in that document and to suggest improvements for TAP 2002–2017.

ASSESSMENT OF TAP 00–15

TAP 00–15 incorporated the Mission Area (MA) concept as a way of gauging current and future demand for Army capabilities. Arroyo's analysis of the Mission Areas identified six problems:

- Mixing of operational and institutional functions
- Overlapping areas
- Overly broad Mission Areas
- Inappropriate Mission Areas
- Unwieldy structure
- Imprecise performance measures

Mixing of operational and institutional functions. Since Mission Areas focused on the Army's operational functions, they should not contain tasks associated with institutional functions such as training or maintaining specific types of capabilities, e.g., decontamination. However, MAs in TAP 00–15 contained such functions.

Overlapping areas. Some Mission Areas overlapped with others. For example, the *Perform Other Missions* MA contained missions that belong in the *Peacetime Operations* MA.

Overly broad Mission Areas. The MA *Generate the Force* not only contained institutional functions, it contained almost all of them. Again, the focus of Mission Areas should be on specific operational capabilities so they can facilitate identification of shortfalls.

Inappropriate Mission Areas. TAP 00–15 included *Maintain Force Readiness* as a MA. But no accepted definition of readiness exists. What consensus does exist views it as an output that represents the synthesis of a number of inputs, e.g., training and equipping. Thus, it cut across all MAs and does not provide a good basis for determining specific operational shortfalls to be addressed in the resourcing process.

Unwieldy organizational structure. TAP 00–15 contained 10 MAs further subdivided into 122 operational tasks, which were further subdivided into operational capabilities. There were too many tasks to provide a coherent assessment of what capabilities the Army needs.

Imprecise performance standards. Performance standards are the tools the Army uses to determine its shortfalls in operational capabilities. They measure outcomes or outputs. The General Accounting Office has derived a set of characteristics for good performance standards. When these are applied to those in TAP 00–15, several discrepancies appear. For example, the General Accounting Office (GAO) recommended that performance standards be limited to the vital few. TAP 00–15 has 1,248.

REVISING TAP 02–17

The methodology used to make recommendations for revising TAP 02–17 considered three approaches that refined the MAs and subordinate levels. These three approaches were developed in response to the sponsor's guidance.

Approach 1: refining the current baseline. This approach preserved definitions and existing MAs. It focused on clarifying categories and capabilities.

Approach 2: changing the structure of the current baseline. This approach changed definitions and MAs, and it refined the placement of tasks and capabilities.

Approach 3: creating a hybrid derived from the existing baseline. The last approach used the best of the baseline and Approaches 1 and 2 to form a suggested hierarchy. It is a proposed hierarchy of MAs, operational objectives, and operational capabilities.

These approaches were then assessed using criteria developed in the analysis of the 00–15 TAP. This assessment provided an array of alternative approaches for the Army and assessed their comparative advantages and disadvantages.

RECOMMENDATIONS FOR TAP 02–17

With respect to TAP 02–17, the research team recommends that the Army take the following actions as an outgrowth of Approach 3:

- Reduce MAs to those that focus on operational missions and realign the MA hierarchy—use operational objectives and capabilities as subordinate levels of the MAs. The MAs are dynamic and should be drawn directly from the strategy in the Defense Planning Guidance.

- Commence MA assessments before the publication of Army Strategic Planning Guidance and the beginning of TAP work. The MA assessments should start with the current POM to evaluate what was resourced in the current program.

- Consider placing responsibility for the MAs and assessments (but not TAP) in a different section of the office of the Army Deputy Chief Staff for Operations and Plans (DCSOPS) and give it greater linkage to strategic planning.

- Separate the ASPG, TAP, and APGM from being developed concurrently and insure that they are done in a progressive sequence over a longer period.

ACKNOWLEDGMENTS

The authors acknowledge the following people for their assistance and support: Major Christine Anderson, formally of Resource Analysis and Integration Office, DCSOPS (DAMO-ZR) and now in the Office of Congressional Liaison (OCLL). She oversaw the project and allowed us to explore various options for improving the MA assessments. We would also like to thank Colonel Patrick Bennett (USA, ret.) and Colonel Charles Rash (USA, ret.), who worked with us and members of the Army Staff in developing our analyses.

From RAND we would like to thank our research assistants: Anissa Thompson, Kenneth Myers, Traci Williams, Omid Fattahi, and Matthew Gershwin. Jerry Sollinger provided invaluable recommendations on how to make the manuscript more readable. Deanna Weber edited and assembled the final paper.

The authors, of course, are responsible for any shortcomings in the research.

ABBREVIATIONS

APGM	Army Programming Guidance Memorandum
ASPG	Army Strategic Planning Guidance
AV2010	Army Vision 2010
AVCSA	Assistant Vice Chief of Staff of the Army
CINC	Commander in Chief
CJCS	Chairman of the Joint Chiefs of Staff
CONUS	Continental United States
CPR	Chairman's Program Recommendation
DPA&E	Director, Program Analysis and Evaluation
DCSOPS	Deputy Chief of Staff for Operations and Plans
DoD	Department of Defense
DPG	Defense Planning Guidance
FY	Fiscal Year
FYDP	Future Years Defense Program
GAO	General Accounting Office
GPRA	Government Performance and Results Act
JCS	Joint Chiefs of Staff
JPD	Joint Planning Document

JROC	Joint Requirements Oversight Council
JS	Joint Staff
JV2010	Joint Vision 2010
JVIMP	Joint Vision Implementation Master Plan
JWCA	Joint Warfighting Capabilities Assessment
LOC	Line of Communication
MA	Mission Area
MDEP	Management Decision Package
MTW	Major Theater War
NMS	National Military Strategy
OBP	Objective Based Planning
OCONUS	Outside the Continental United States
OSD	Office of the Secretary of Defense
PEG	Program Evaluation Group
POM	Program Objective Memorandum
PPBES	Planning, Programming, Budgeting, and Execution System
QDR	Quadrennial Defense Review
SSC	Smaller-Scale Contingencies
TAP	The Army Plan
TRADOC	Training and Doctrine Command
UJTL	Universal Joint Task List
WMD	Weapons of Mass Destruction

Chapter One

INTRODUCTION

BACKGROUND

During fiscal years (FY) 1995 and 1996, the Army reengineered its programming process. The reengineering required both organizational and procedural changes.[1] For example, a new position, the Assistant Vice Chief of Staff Army (AVCSA), was created and given the responsibility for program integration across the Army Staff. The Program Evaluation Groups (PEGs), which play a central role in programming, were reorganized, and Mission Area (MA) teams were established within the office of the Deputy Chief of Staff for Operations and Plans (DCSOPS).

PURPOSE OF THIS PROJECT

In 1997 the Army asked that RAND Arroyo Center assist in the reengineering of The Army Plan (TAP). The Army also reinstituted strategic planning and the product was the Army Strategic Planning Guidance (ASPG), which provides strategic guidance to TAP.[2] Planning and programming in the Army are centralized within the Department of the Army's headquarters. This document discusses the Arroyo Center's work on TAP.

---

[1]See Leslie Lewis, Roger Brown, and John Schrader, *Improving Army PPBES: The Programming Phase*, Santa Monica, CA: RAND, MR-934-A, 1999, and Leslie Lewis, Roger Brown, and John Schrader, *Improving the Army's Resource Decisionmaking*, Santa Monica, CA: RAND, DB-294-A, 2000.

[2]Leslie Lewis, Roger Brown, and John Schrader, unpublished research.

The Army had substantially changed its programming process and associated documents. The Arroyo Center was asked to assess several documents, including TAP for the years 2000–2015, to determine to what extent the reengineering that had been done was successful and suggest improvements for the next TAP (2002–2017). This report documents RAND's assessment and recommendations. The Army specifically asked for comment on the following areas:

- Independence of MAs and PEGs

- Focus of MAs

- Operational objectives

- Performance measures, standards, and risk assessment

RAND's analysis of the other Army planning and programming documents is reported elsewhere.[3]

The Army is continuing its reengineering efforts and implementing further changes.

THE ARMY PLAN (TAP)

TAP, like other planning and programming documents, is published biennially in the odd years. It reflects the National Military Strategy (NMS) and the Defense Planning Guidance (DPG) and guides the Army's Planning, Programming, Budgeting, and Execution System (PPBES).[4] It draws on planning scenarios to identify combat force requirements and, for each program year, develops a force that meets the requirements within anticipated personnel and budget ceilings. TAP projects requirements for the near term (0–5 years), middle term (6–15 years), and long term (15–20 years). TAP is designed to set the initial programming priorities within the anticipated resources. TAP contains three main sections. The first provides the Army's strategic planning guidance. The second articulates institutional goals and objectives, and the final section is the APGM.

---

[3]See Lewis et al., *Improving the Army PPBES: The Programming Phase.*

[4]Elements of this description have been drawn from *How the Army Runs: A Senior Leader Reference Handbook, 1997–1998,* Carlisle Barracks, PA: U.S. Department of the Army, 1997.

HOW THIS REPORT IS ORGANIZED

This report has five chapters. Chapter Two describes the context in which TAP is developed. Chapter Three presents those portions of work on reengineering Army programming that are relevant to TAP. Chapter Four contains the analysis of the 00–15 TAP. Chapter Five presents recommendations about how the 02–17 TAP should be modified, and Chapter Six provides a brief summary and conclusions.

THE CONTEXT OF TAP

This chapter sets TAP in the context of the external and internal processes and documents that it interacts with and responds to. TAP is not produced in a vacuum, and it is important to understand how other processes and documents affect it because they shape much of the material TAP incorporates. In turn, TAP also sets in motion several internal Army processes and reacts to others. TAP is the primary planning document that provides the Army with planning and programming guidance. The responsibility for TAP comes under the Chief of the Resource Analysis and Integration Office.

To understand the dynamics of the supply, demand, and integration model and how it provides for recognition and adjudication of these different demands, it is necessary to understand the relationships among key external and Army processes and products.[1] Equally important is the time horizon along which each of the identified elements operates.

Figure 2.1 portrays these elements. The time horizons in the figure are near (now to five years out), near-to-mid term (6–15 years out), and the future (15–20 years out).

The left-hand side of Figure 2.1 identifies those documents and processes associated with the Office of the Secretary of Defense (OSD) and the Joint Staff (JS) that affect both Army planning and pro

---

[1]See Figure 3.1 in Chapter Three for further explanation of the supply, demand, integration model.

RAND *MR1133-2.1*

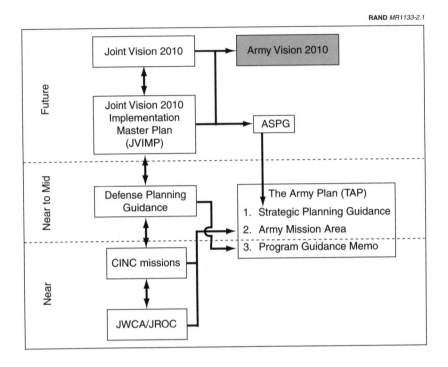

Figure 2.1—Key TAP Relationships

gramming. TAP is shaped by both documents and processes. Two documents, the Joint Vision 2010 (JV2010) and Joint Vision 2010 Implementation Master Plan (JVIMP), were developed by the JS and have ramifications for how external demands shape Army operational and, to a lesser extent, institutional demands.

The initial version of JV2010 was published in 1996. JV2010 is operationally focused on future ways to fight. It argues for a full-spectrum joint force that performs seamless operations and defeats opponents by dominating information, by both providing it to friendly forces and denying it to opponents. The adoption of new operational concepts through the application of leap-ahead technologies is critical to JV2010. For example, focused logistics, one of four central operational concepts featured in the document, argues that technology can reduce the large logistics tails that typified operations in World

War II and the Gulf War. Rapid transportation technologies to provide assets quickly and information technology to track and shift assets as needed would reduce logistics requirements.[2]

Soon after the publication of JV2010, the JS initiated an examination of how it might be implemented. JVIMP builds on the general concepts laid out in JV2010 and identifies potential paths for how the concepts might be developed and implemented. It concentrates on the application of technology in order to attain full-spectrum dominance, that is, dominance in all types of conflict.[3] The critical element of JVIMP is its focus on the services' responsibilities for providing joint capabilities and the required coordinating activities.

The DPG provides guidance to both the operational and institutional elements, each of which must be specifically addressed in the Army's program. The Joint Warfighting Capabilities Assessment (JWCA) process provides a series of JS assessments of the need for and availability of current and future operational capabilities, as defined by the Commanders in Chief (CINCs). The JS then uses the output of the various JWCA assessments to evaluate the services' ability to meet CINC demands now and in the middle term. The operational demands established by the CINCs and refined in the JWCA/Joint Requirements Oversight Council (JROC) process are linked to the Army through the Army Mission Areas.

The right side of Figure 2.1 begins with Army Vision 2010 (AV2010), which was developed by the Army in response to JV2010. Its publication preceded that of the JVIMP. Like JV2010, AV2010 builds on the concept of information dominance but within the Army's operational construct. For instance, information dominance is accomplished through a concept of mental agility that calls for digitization of the Army, including many of its existing platforms. AV2010 focuses on major theater wars (MTWs). Operations such as peacekeeping and humanitarian assistance are treated as lesser included activities; if the Army can handle MTWs, then the capabilities required to conduct smaller-scale contingencies (SSCs) and peace-

---

[2]General John Shalikashvili, *Joint Vision 2010*, Washington, D.C., 1996, pp. 10–11.

[3]*Joint Vision Implementation Master Plan*, Washington, D.C., June 1, 1997.

keeping-like operations are accommodated.[4] This is defined as full-spectrum dominance. Appendix A illustrates the linkages among JV2010, AV2010, and the ASPG. It also illustrates the operational and institutional demands placed upon the Army.

As Figure 2.1 suggests, all of these documents and processes help shape TAP. TAP provides planning guidance to the Army based on external processes. TAP is also the primary planning document that provides programming guidance to the Army. Importantly, any proposed changes to Army processes must accommodate these external processes and functions. The Army's program must demonstrate that it is responsive to the diverse issues raised by OSD, the Joint Staff, and Congress. For example, it would be difficult to discuss the Army's operational readiness and justify resource choices to the CJCS by exclusively discussing readiness within the context of Army modernization. On the other hand, it would be difficult to describe to the OSD and Congress how the Army is caring for its military families by a discussion of its operational readiness, although if dependents are not being sufficiently taken care of, overall Army readiness would most certainly be affected.

---

[4]This assumption was later repudiated by parts of the Army during the Quadrennial Defense Review. Army analyses showed that SSCs and peacekeeping-type operations often required MTW-like capabilities as well as sets of capabilities that are mission unique. For example, peacekeeping operations often require large numbers of military police, a capability that is also required in MTWs but perhaps in smaller numbers. Most peacekeeping operations also require foreign area officers who are language proficient and knowledgeable about local customs and behaviors, a capability not often required in large-scale ground operations with two major opposing forces.

Chapter Three

REENGINEERING ARMY PROGRAMMING

This chapter describes the aspects of the reengineering that are most important to TAP and shows the mechanisms the Army established to implement them. A model of supply, demand, and integration is central to the Army's reengineering efforts and to what should appear in TAP 02–17. This chapter and the previous one are preconditions to understanding Chapter Four, the analysis of the results of the reengineering of the 00–15 TAP.

THE RAND FRAMEWORK

This assessment of TAP is shaped by a framework derived from the economic model. The RAND-developed framework appears in Figure 3.1.

The operational demand, depicted on the left side of the diagram as joint demand for Army capabilities, may be thought of as demand in economic terms. It consists of those capabilities that only the Army can provide to the CINCs to perform their missions. The institutional demands, depicted on the right side of the figure as capabilities provided by the Army, may be thought of as the supply and are those activities that support the generation of the Army-unique capabilities needed to satisfy the CINCs' operational demands. Put in economic terms, they are the "supply" that meets the CINCs' demands. For example, "train future leaders" is an institutional requirement, as is "ensure that the Army's research and technology programs are sufficiently funded to provide needed information

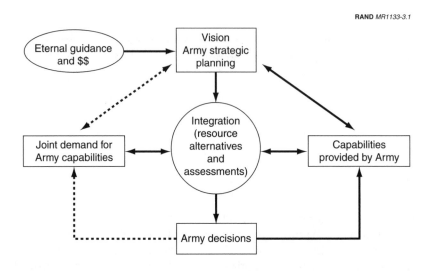

RAND *MR1133-3.1*

Figure 3.1—Programming Framework

technology in support of Army XXI." The arrows indicate the iterative nature of the process.[1]

THE ARMY RESPONSE

Figure 3.2 shows the framework implemented within the Army as of spring 1998 and its relationship to the various external organizations and functions. The figure reiterates that there is a single Army in which two distinct elements must be recognized and resourced: the institutional, indicated by the Army Title 10 functions shown on the right, and the operational, indicated by the Army MAs on the left. This concept is expanded upon to accommodate how the external environment shapes Army demands through existing processes and functions. Figure 3.2 shows these relationships and their interactions with the Army.

---

[1]In another context, the MA/PEG relationship is analogous to supply and demand. The PEGs are on the supply side, while the MAs are on the demand side. They must be kept separate in order to have functions independent of each other.

The left-hand side of the figure shows the linkage between the CINC warfighting missions and Army MAs. The CINC missions are extracted from the DPG, the Quadrennial Defense Review (QDR) report, and the recently published Department of Defense (DoD) guidance on new missions for the DoD. The Government Performance Plan for FY99 also lists the DoD MAs. The structure assists the Army in translating the DoD MAs into Army MAs. The derived Army MAs were created to facilitate the identification and assessment of progress toward operational objectives, tasks, and related capabilities that are closely associated with the Army. The Army MAs enable the Army to focus on those capabilities most closely associated with it but within the broader context of the joint environment.

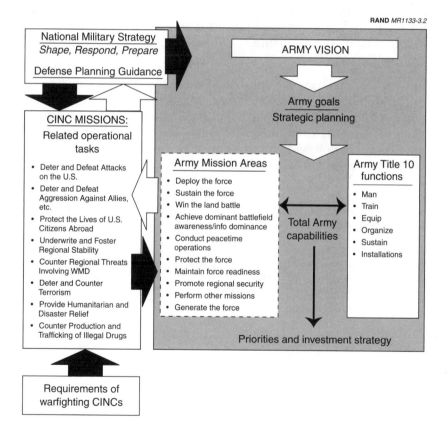

Figure 3.2—Relating Army Capabilities to CINC Missions

Two aspects of the figure represent a marked departure from past procedures. Most notable is the creation of the Army MAs as a way of ensuring that planning and programming support the CINC requirements. The second change is the organization of the PEGs along the lines of the six broad Title 10 functions.[2] The function of PEGs is to allocate Army resources to programs over the POM/FYDP years. The PEGs follow priorities set in the APGM as they allocate resources. The composite effort of the PEGs results in decisions by the Army leadership on the Army POM. Previously, the Army had 14 PEGs organized to support a combination of staff functions, organizations, and special interest areas that worked independently. Little coordination occurred among them; the resulting Army program represented an aggregation of individual program groups rather than a truly integrated program. Furthermore, without the Army MAs, there was no way to ensure that the Army program satisfied the demands of the CINCs.

In FY97–98, the Army began to implement the operational element as part of its reengineering activities. MAs, which focused on identification and evaluation of those Army capabilities that were critical to supporting the CINC missions, were introduced. The Army implementation, however, fell short of producing a useful mechanism to determine needed capabilities.[3]

---

[2]Title 10 spells out twelve functions for the Army. However, only six have direct resourcing implications and therefore, only six PEGs were created: man, equip, train, sustain, organize, and installations.

[3]The intent of the concept was to provide assessments of current and future capabilities that Army programming would resource through the efforts of the PEGs. For further explanation of the PEG/MA relationship, see Leslie Lewis, Roger Brown, and John Schrader, *Improving the Army's Resource Decisionmaking*, Santa Monica, CA: RAND, DB-294-A, 2000.

Chapter Four

ANALYSIS OF TAP 00–15: WHAT WAS ACCOMPLISHED

The Army Plan was developed under challenging conditions. The development of the ASPG, which becomes the new first section of TAP, represented the first time such an effort had been carried out in several years. In prior years, the initial sections of TAP provided summaries of the strategic environment, potential threats, and implications for the Army, along with Army leadership strategic guidance for the future. This was broad in nature and lacked the specific objectives, transformation path, and connectivity now found in ASPG.[1] (Appendix B explains how the guidance in the ASPG translates into TAP and how the Army subsequently responds to the guidance.) Further, the schedule for POM 00–05 was compressed. The result was a marginally changed TAP 00–15 in which the three major sections were prepared concurrently rather than in the intended sequence over a longer period.

ARMY MISSION AREAS (TAP 00–15)

This analysis focuses on the new section in the reengineered TAP called the Army Mission Areas, which is section 2 of TAP. The MA concept was designed as a method by which the Army could assess the current and future demand for operational capabilities.

MAs provide greater specificity than is found in the Army vision and strategic planning guidance about how the Army will meet the strategic demands created by national strategies, defense guidance,

---

[1]Leslie Lewis, Roger Brown, and John Schrader, unpublished research.

and broad and continuing CINC missions through the Army's vision and core competency. MAs aggregate the Army's operational capabilities in an understandable way. As such, MAs are the broadest aggregates of an Army-level hierarchy that proceeds down through operational tasks and capabilities. At the next-highest level of the hierarchy, operational tasks are not as specific as those that exist at the lower operational capability or performance standard levels of the hierarchy. However, MAs need to be precise enough to present the Army in "what-we-can-do" terms. MAs and their subordinate components are the Army to the nation, to Congress, to the OSD, and to the CINCs. MAs should be expressed in a manner that allows an assessment to be made of how well each is being achieved.

MAs come from a relentless focus (through the lens of the Army vision and core competencies) on the external operational requirements.[2] This focus should be on current and future demands, not on what has historically been provided. MAs should be distinct from each other; they should be mutually exclusive to gain clarity and to allow statement of clear priorities among them. The MAs cut horizontally across the institutional functions, and each MA must have an advocate with a broad operational or doctrinal perspective on how the internal functions enable the operational capabilities. MAs guide decisions about courses of action and about the allocation of inputs by internal functions to obtain desired outputs and outcomes. Investments in these institutional functions, rather than in MAs directly, set future direction and allow assessment of the Army's ability to provide needed operational capabilities.

MAs are more short-lived than the more enduring Army Title 10 missions, Army vision, and Army core competency. MAs, operational tasks, and operational capabilities are something to be accomplished

---

[2]The U.S. Army Vision describes the world's best army, a full-spectrum force trained and ready for victory. It is a total force of quality soldiers and civilians: a values-based organization, an integral part of the joint team, equipped with the most modern weapons and equipment that the country can provide, able to respond to our nation's needs, and changing to meet challenges of today, tomorrow, and the 21st century. The Army's core competency is composed of quality people, leader development, force mix, training, doctrine, and modern equipment. *Army Strategic Planning Guidance '99*, Draft, March 2, 1999. Subsequently, Army Chief of Staff Erik Shinseki released a new Army Vision on October 12, 1999.

at particular moments over the planning horizon. MAs may change as strategic demands change.

Where Do Mission Areas Come From?

The MAs derive from the CINC missions, and they are specified in the ASPG. Analysis of the CINC missions determined that they were too broad for Army programming purposes, and resulted in an initial set of MAs, identified using a taxonomy called Objectives-Based Planning (OBP).[3] Work in TAP 00–15 concentrated on the introduction of the MAs and some initial assessments of the sufficiency of current and future Army capabilities by ad hoc groups. This latter effort failed to provide a useful methodology for assessment.

What Are the Army Mission Areas?

The MAs reflect broad operational activities that the Army has to perform. Table 4.1 lists the 10 Army MAs contained in TAP 00–15.

Table 4.1

Mission Areas for TAP 00–15

Win the Land Battle
Conduct Peacetime Operations
Promote Regional Stability
Perform Other Missions
Achieve Dominant Battlefield Awareness and Information Dominance
Deploy the Force
Protect the Force
Sustain the Force
Generate the Force
Maintain Force Readiness

NOTE: The Army added three Mission Areas to the seven that the Arroyo Center developed: Deploy the Force, Generate the Force, and Maintain Force Readiness.

---

[3]The OBP methodology has a long history at RAND. Initially developed for use by the Air Force, it was called Strategy-to-Tasks. The methodology was later expanded to link to specific resources and to be much more focused on joint operations. See Kent (1983), Lewis and Roll (1993), Pirnie (1996), and Lewis, Pirnie, et al. (1999).

How Are Mission Areas Organized in TAP 00–15?

Each MA was organized into a series of operational tasks that are the primary things the Army has to do in the particular MA.[4] Tasks in turn are composed of operational capabilities that enable the Army to accomplish these tasks. Each MA also has performance standards to help the Army determine how much is required under optimal circumstances and, alternatively, how much is critical to accomplish the task with acceptable risk. An example, drawn from TAP 00–15, of one MA and its subcomponents appears below.

Mission Area:	Win the Land Battle
• **Operational task:**	Maintain rear area security
• **Operational capability:**	Provide forces and equipment to secure lines of communication (LOC)
• **Performance standard:**	Ensure sufficient forces and equipment to secure designated LOC in accordance with operational requirements

TAP 00–15 contained 10 MAs, 122 operational tasks, 789 operational capabilities, and 1,248 performance standards.[5]

How Were Mission Areas Found in TAP 00–15 Evaluated?

To assess the MAs, tasks, and capabilities, a set of criteria frequently were applied to statements of objectives in decisionmaking. Objectives should have the following characteristics:[6]

- **Operational:** done without functional or institutional elements.

- **Reasonably complete:** balanced and adequately addressing all needs and concerns.

---

[4]For TAP 00–15, the organizational structure for MAs consisted of Operational Tasks, Operational Capabilities, and Performance Standards. See *The Army Plan, FY 2000–2015*, 1998.

[5]U.S. Army, *The Army Plan FY 2000–2015*, 1998.

[6]See for example, Ralph L. Keeney, "Structuring Objectives for Problems of Public Interest," *Operations Research*, Vol. 36, No. 3, May–June 1988, p. 396, or Kraig W. Kirkwood, *Strategic Decision Making*, Belmont, CA: Duxbury Press, 1997.

- **Unique, not redundant:** importance within one part of the hierarchy should not depend on importance in other parts of the hierarchy.

- **Operable and measurable:** must be logically tied to higher and lower levels of the hierarchy and convey measurable expectations.

- **Few in number:** must be understandable and explicable.

- **Performance standards:** assessibility of these standards.

To assess performance standards, we drew on characteristics that the GAO has determined such standards should meet:[7]

- Show degree to which desired results are achieved

- Limited to a vital few, essential for decision making

- Responsive to multiple priorities; balance competing demands

- Responsibility linked; establish accountability for results

What Problems Did We Find With Mission Areas?

We identified six problems with the set of MAs in TAP 00–15: some confuse operational and institutional functions, some overlap with other MAs, some are too broad, some are inappropriate, the current organizational structure of the individual MAs is unwieldy, and the performance standards are neither useful nor usable.

Mixing operational and institutional functions. Some of the current MAs confused operational and functional activities. For instance, the Generate the Force MA replicated a number of tasks associated with the Title 10 functions, such as provide sufficient forces. Furthermore, the MAs contained such functional tasks as "Assure the U.S. ability to operate in chemically and biologically contaminated areas" (Army Mission Area Data Base, January 15, 1998, task ab1) and "Maintain a system for evaluating the Reserve Component deployment prerequisites" (Army Mission Area Data Base, January 15, 1998,

---

[7]U.S. GAO, *Executive Guide: Effectively Implementing the GPRA*, GAO-GGD-96-118, Washington D.C., June 1996.

task cp4), both of which are functional activities associated with the institutional Army. An example of a good operational task is "control movement with roadblocks." Therefore, this mix of operational and institutional functions does not meet our "operational" criterion.

Replicating PEG activities diminishes the purpose of the MAs, which need to focus on the identification of shortfalls in operational capabilities. Attempts to link the MAs to PEG functions through identification of those functions does not inform the PEGs of operational requirements nor what corrective actions are necessary by the PEGs to fix identified operational shortfalls.

Overlapping Mission Areas. Arroyo research provided a baseline list of MAs to the Army. The project identified the Perform Other Missions MA as a collector for such missions as counterdrugs and counterterrorism that could not be logically grouped within the other MAs. However, subsequent Army work identified a new MA: Conduct Peacetime Operations. Since this new category subsumed such missions as counterdrug, it should now become the collector for all the missions identified in Perform Other Missions, which should thus be abolished, its operational elements grouped under Conduct Peacetime Operations.

As an additional note, Perform Other Missions also included a number of institutional tasks, e.g., to identify training issues. Again, these should be removed from the MA because they are PEG tasks that are replicated in the training, equipping, and manning PEGs. This does not meet the "unique, not redundant" criterion.

Overly broad Mission Areas. The MA Generate the Force is the most problematic, for not only does it include PEG functions, it included most of them. Additionally, it does not capture all the institutional elements necessary to generate the force and therefore does not meet the "few in number" criterion. This MA reflected the attempts by the mission area teams to direct PEG activities by linking to all Army resources (like the PEG), rather than to inform the PEGs about the operational shortfalls across the POM that need to be addressed. The MA also lost the focus on informing the external audiences about what operational capabilities the Army provides to CINCs and Army emphasis on directing the PEGs' decisions.

Inappropriate Mission Areas. Figure 3.2 lists one MA as Maintain Force Readiness. Readiness is a complex issue; no consistent definition exists within the DoD. But since the QDR, readiness has increasingly been defined as an output of resourcing activities, rather than as an input. It is viewed as consisting of all the elements necessary to produce a force capable of providing operational capabilities to the CINCs. Thus, readiness is measured as the output of everything a service does to develop and sustain a force: training, equipping, manning, etc.

Defining readiness as a MA is problematic because it limits its utility to only focusing on the PEG activities; it also replicated both the Generate the Force and Sustain the Force MAs, which in turn also partially replicate PEG tasks. Readiness cuts across all the MAs and the PEGs, for in its broadest definition, it provides the ability to measure how a service has resourced to meet its near-, mid-, and long-term operational requirements. This fails to meet the "performance standards" criterion because readiness is an output function.

Unwieldy organizational structure. Each MA has a number of tasks. TAP 00–15 had 122 distinct operational tasks across the 10 MAs. The current structure relies on operational tasks as a way to identify required operational capabilities. From a planning and programming perspective, there were too many operational tasks to provide a coherent assessment to the Army leadership and external audiences of what operational capabilities the Army needs. Furthermore, operational tasks were not unique to a particular MA; they were repeated across many. The resulting operational capabilities could not be assessed based on their ability to perform the totality of the MA; they can only be evaluated against a particular task associated with multiple MAs' attempts to direct the PEGs.

This approach resulted in a determination of the requirement for a particular operational capability, based on how often an operational task was repeated across all the MAs. Having MAs that are focused on Title 10 functions compounded the problem, and therefore much of the analysis resulted in operational capabilities that are defined solely by their Title 10 elements of people, equipment, infrastructure, and training. Thus, Title 10 requirements were identified, and the operational capability was rarely if ever identified. This does not meet the "operational" and "few in number" criteria.

Imprecise performance standards. The implementation of the Government Performance and Results Act (GPRA) placed increasing emphasis on measurement, particularly the measurement of results. Performance measurement that is tangible and objective is deemed best. Such measurement incorporates several aspects. First, a metric or scale represents the specific characteristic used to gauge performance. This can be thought of as a yardstick (feet, inches) if the characteristic of interest is length. In essence, the scale or metric has to relate to what is being measured. The second part is a goal or *standard*, which is the target level of performance to be achieved. For example, one might want to cut two-by-fours into three-foot lengths. The performance goal is three feet on the yardstick metric. The third part is a *measure* of actual achievement. One can assess the results of cutting the two-by-fours. Within some tolerance, the activity of cutting has met the specification (standard) of three-foot lengths or it has not.

The performance standard is what one is trying to achieve with a plan, program, or activity, and this can be measured (on the relevant scale) against actual results. Results can be of two types: outcomes and outputs. Outcomes are the final results achieved in relation to desired objectives. Outputs are the intermediate results of processes and contribute to outcomes. For example, units (organized people and equipment) are an output that contributes to a final outcome. Outcome measures are better than just measures of outputs, but they are not always possible. Conversely, some systems are so complex that neither output nor outcome measurements seem possible, so inputs to the processes are measured instead. Moreover, GPRA also recognizes that some systems do not easily allow for objective or tangible measurement. When objective measurement is not possible, subjective assessments may be made but must be in terms that would permit an independent determination of whether the eventual performance corresponded to the desired performance statement.

The GAO has stated that performance standards should have four characteristics:

- They should be tied to objectives (desired outcomes) and show the degree to which desired results are achieved.

- They should be limited to a vital few essential for decision-making.

- They should be responsive to multiple priorities in order to show balance across competing demands (e.g., cost and quality).

- They should be in order of responsibility linked to establish accountability for results.

Given performance standards that meet these characteristics, an organization can collect data (using the relevant metric or scale) that are sufficiently complete, accurate, and consistent to be useful.

Performance standards contained in TAP 00–15 were flawed when judged by these characteristics and, consequently, were neither useful nor usable. Examples of some of these standards are statements such as "provide personnel, equipment, materiel" or "provide resources." These very general statements are not tied to outcomes, nor do they show the degree to which the desired results (outcomes, outputs) are achieved. They may or may not show the degree to which inputs are provided. There are 1,248 performance standards in TAP 00–15, which certainly appears to be more than the vital few. It is not clear that they respond to multiple priorities or have responsibilities linked to them.

Defining performance standards that meet the characteristics and allow for complete, accurate, and consistent data (using the relevant metric or scale) to be collected to assess results is easier said than done. The next chapter offers some suggestions to guide continuing efforts in this area.

Initial Resourcing Guidance: APGM and Its Use

There were few problems with the section of TAP that provides the initial resourcing guidance. Title 10, or institutional, functions were very well articulated and understood, as would be expected given that the Army has employed this approach for some time.

Some action officers on the Army Staff wanted to merge the Army Program Guidance Memorandum (APGM) with TAP. TAP provides the initial programming guidance that lays out the key Army planning objectives (Section 1) and the demand for operational capabilities (Section 2). Section 3 should provide the initial institutional or PEG guidance. Historically covering PEG goals, resource objectives, tasks, and priorities, the APGM has been a separate document that contains the final programming guidance that immediately proceeds the beginning of the program (POM) build. It is not considered part of TAP.

The agreement was reached in the development of TAP 00–15 that its Section 3 would contain the initial detailed PEG guidance and that a subsequent APGM would provide the final programming guidance. Thus, Section 3 and the APGM were to be called APGM I (Section 3 of TAP) and APGM II (Final Programming Guidance) respectively.

Chapter Five

REVISING TAP 02–17

This chapter describes three approaches for modifying TAP 02–17 and our assessment of the approaches in conformance with the sponsor's guidance. It explores several approaches for refining the MAs and subordinate levels for TAP 02–17. The first part of this chapter describes three approaches to refining the MAs: (1) Refine the current baseline (preserve definitions and existing MAs; make the structure more hierarchical and refine the placement of tasks and capabilities. (2) Change the structure of the current baseline (change definitions and MAs; make hierarchical and refine the placement of tasks and capabilities). (3) Create a hybrid (using insights gained from the first two approaches). These are followed by suggestions about performance measures; we treated these separately because they could apply to any of the three approaches.

APPROACH 1: REFINE THE CURRENT BASELINE

The first approach proceeded MA by MA through all 10 in the existing database. We moved tasks that did not seem to fit the MA; we eliminated redundant operational tasks and their operational capabilities; we demoted and eliminated subordinate operational tasks and then eliminated remaining redundant operational capabilities. We did not change definitions, rewrite task or capability statements, aggregate (group several into one higher level), disaggregate (break one into several), or consolidate (merge into one) any elements. We did not check tasks for completeness against MAs, nor did we check capabilities for completeness or operability against tasks. Rather we focused on clarifying categories and their functions. Table 5.1 provides an example of how we refined the hierarchy in one MA, Conduct Peacetime Operations.

After refining the MAs, we assessed the results against the criteria discussed in Chapter Four. Ten MAs were still too many and of the wrong type for explaining the Army to external audiences. Nor were they useful as collectors of operational capabilities. The operational tasks and capabilities were generally not complete in that they lacked a doctrinal basis and needed better statements to be more understandable and useful. Moreover, additional and different operational tasks and operational capabilities were needed. It appeared that some tasks could be elevated or aggregated into a higher level. We pruned the number of tasks from 122 to 74 and the number of operational capabilities from 789 to 467, but the numbers were still too large. We believe we had eliminated redundancy among the operational tasks and capabilities throughout the hierarchy, making it somewhat more operable. However, they were still not measurable in that there were few objective and measurable expectations inherent in them.

Table 5.1

Refining Conduct Peacetime Operations Mission Area

Operational Tasks	Action
• Reduce will of opponent to fight	Keep
• Construct, maintain, or repair required infrastructure	Keep
• Control movement within and across borders outside continental United States (OCONUS)	Keep
• Assist in maintaining civil order	Keep
• Support activities of non-governmental organizations	Keep
• Employ total force	Eliminate (redundant)
• Secure electoral activities	Subordinate
• Conduct post-hostility operations	Move to other MA
• Suppress or destroy opposing air defenses and command, control, communications, computer, and intelligence (C4I)	Move to other MA
• Acquire and disseminate intelligence about opposing force	Move to other MA

APPROACH 2: CHANGE THE STRUCTURE OF THE CURRENT BASELINE

The second approach had three components. First, we revised the taxonomy of the MA by elevating operational capabilities above the task level and by adding a new level between the operational capability and the MA, the operational objective. We provided to the Army examples of operational objectives. We then crafted new definitions for a MA, operational capability, and operational task. A revised taxonomy appears below.

Mission Area:	Win the Land Battle
— Operational objective:	Dominate enemy forces in theater
• Operational capability:	Provide forces and equipment to secure LOC
— Operational task:	Maintain rear area security

Since we had created a new level and altered the subordination of the old one by switching operational tasks and capabilities, we prepared a new set of definitions for each level of the hierarchy. These are as follows:

Component of Mission Area	What It Does
Mission Area	Explains why the Army exists and tells what it does. Outlines what the Army must do to support CINC missions, and collects operational objectives.
Operational objective	Describes in a **goal** or **objective** statement of what the operational Army must do to meet CINC needs. Groups operational capabilities logically.
Operational capability	Defines broad operational **activities** the Army has to perform to accomplish operational objectives.
Operational task	Provides the detailed **actions** that are necessary to produce the operational capability. Provides specific units of effort that can be measured in terms of time cost and throughput. Answers the what, when, where, and how-much questions and provides the requirements for resources.

Armed with this hierarchy and new set of definitions, we provided examples of operational objectives. For example, under the MA Win the Land Battle, we added the operational objectives of "dominate enemy forces in theater," "force entry into theater," and "degrade opposing stocks and infrastructure."

We also reassessed the MAs, eliminating those that did not pertain directly to operational missions and eliminating or moving their associated operational tasks. For example, under the MA Win the Land Battle, we eliminated such tasks as "conduct civil military operations," "conduct opposed amphibious landings," and "conduct opposed airborne assaults." We moved capabilities that did not fit with operational objectives and consolidated or eliminated subordinate or redundant capabilities and tasks. Finally, we demoted operational tasks within the hierarchy to fit under the remaining operational capabilities.

However, in this approach we did not create the full level of operational objectives nor rewrite operational capability or task statements. We did not aggregate or disaggregate. We did not check capabilities for completeness against MAs or check tasks for completeness or operability against capabilities. We provided new ordering of existing data based on sponsor guidance; therefore we did not provide a new complete taxonomy.

This approach resulted in the reduction of the MAs to seven that appeared to be useful. We had suggested some operational objectives because that level in the proposed hierarchy was missing. The operational tasks and capabilities had the same deficiencies outlined in the first approach with one exception: they were fewer. We had pruned the number to 52 capabilities and 374 tasks.

APPROACH 3: CREATE A HYBRID DERIVED FROM THE EXISTING BASELINE

The last approach was a hybrid. Essentially, we used the best of the baseline and the two approaches above and put together a suggested hierarchy. Several combinations were possible, and we proceeded in an iterative fashion, testing several against the criteria, until one emerged as "best." Best is in quotation marks because it is but a strawman and can be made better by Army experts. We propose a hierarchy of MAs, operational objectives, and operational capabilities as defined above. We also suggest how and when MA analyses might be made as part of the resourcing process. We note up front that the hierarchy we outline is not complete because the levels of objectives and capabilities have not been fully set forth.

Mission Areas

MAs elaborate the implications for the Army in providing operational capabilities to meet joint demands. They provide greater specificity as to how the Army will meet the demands created by national strategies, defense guidance, and broad and continuing CINC missions through the Army's vision and core competency. The MAs should allow assessment of whether each was or is being achieved (GPRA; OMB Circ A-11). MAs and objectives are something to be accomplished at particular moments over the planning horizon.

We derived a suggested list of MAs from national security and defense literature and from statements by Army leaders:[1]

- Promote Regional Stability

- Deter or Reduce Conflicts or Threats

- Fight and Win Major Theater Wars

- Conduct Smaller-Scale Contingency Operations

- Secure the Homeland

- Prepare Forces and Provide Capabilities

- Exploit Concept Innovation and Modernize Forces Accordingly

An audit trail for these MAs is contained in Appendix C.

---

[1]We reorganized the first five MAs from the DPG. The last 2 MAs were added by request from the sponsor to ensure that the MA structure would cover all the resources of the Army. The project did not endorse this request. Subsequently, the following seven MAs were decided upon by the Army in ASPG '99: Promote Regional Stability; Reduce Potential Conflicts and Threats; Deter Aggression and Coercion; Conduct Smaller-Scale Contingency (SSC) Operations; Deploy, Fight, and Win Major Theater Wars (MTWs); Secure the Homeland; and Provide Domestic Support to Civil Authorities.

Operational Objectives

Operational objectives aim at Mission Areas and are the aim of subordinate operational capabilities. Operational objectives guide and stimulate capabilities. They complement a MA in which achievement cannot always be directly or objectively measured. The assessment is made on the objective rather than on the MA.

We drew on several sources for a strawman list of operational objectives, including the Universal Joint Task List (UJTL),[2] the Army Blueprint of the Battlefield,[3] and prior RAND work on identifying operational objectives.[4] We also reviewed various statements in the existing TAP that were candidates to become mission objectives. Some were kept as is; others were aggregated; some were consolidated. To do this, we applied a standard set of criteria to determine an operational objective:

- Is it clear?

- Is it balanced with other objectives?

- Is it goal oriented, i.e., a guide to action?

- Is it explicit enough to suggest certain capabilities?

- Is it suggestive of measurement and control?

- Is it ambitious enough to be challenging?

- Does it suggest cognizance of external and internal constraints?

- Can it be related more broadly (mission areas) and more specifically (operational capabilities and tasks) at higher and lower levels in the organization?

---

[2]U.S. Department of Defense, *Universal Joint Task List,* Washington, D.C., 1995.

[3]"Army Blueprint of the Battlefield" is a concept-developments template that portrays the integration of all the Army Battlefield Operating Systems.

[4]Bruce Pirnie, *An Objectives-Based Approach to Military Campaign Analysis,* Santa Monica, CA: RAND, MR-656-JS, 1996.

Strawman Operational Objectives

We applied the individual objective criteria in building the following list; however, this list is a strawman because it was not tested against the criteria for a full set of objectives outlined earlier. See Appendix D for the rest of the strawman operational objectives.

- Maintain military-to-military contacts

- Provide assistance

- Participate in exercises

- Maintain presence

- Defend and protect U.S. and allied forces

- Conduct show of force and other demonstrations

- Prevent proliferation of WMD and conventional weapons

Operational Capabilities

A capability is an efficient combination of organizations, technology (systems, equipment, processes, services), manpower, and training in support of an objective. These ingredients of a capability are the outputs of the various planning and resource processes. Operational capabilities change—disappear, evolve, emerge—over time (the transition path) as needs and resources arise or diminish.

Operational capabilities meet a similar set of criteria as objectives. Operational capabilities at TAP level are activity statements of desired performance and can be prioritized. More quantitative and measurable statements of performance will be contained in subordinate plans and analyses. Operational capabilities at TAP level of analysis are not stated in terms of specific units, weapon systems, or numbers of people. However, other plans and programs frequently are stated in these more specific terms, and it is the aggregate of these ingredients that informs the mission and capability assessment.

Strawman Operational Capabilities

The strawman operational capabilities are not complete. In particular, objectives have been replicated where subordinate capabilities have not been stated. See Appendix E for the rest of the strawman operational capabilities.

- Maintain military-to-military contacts

- Training assistance

- Materiel assistance

- Participate in exercises

- Station forces OCONUS

- Deploy forces periodically

- WMD protection

- Conventional protection

- Terrorism protection

In Appendix F is a proposed hierarchy. It is not complete because the operational objectives and capabilities have not been fully set forth.

ASSESSMENT OF THE THREE APPROACHES

To understand the ability of each proposed approach to define operational requirements and provide clear ways to measure progress, we evaluated each one against the criteria identified in Chapter Four. Figure 5.1 shows our own assessment of the three approaches.

Approach 1, refinement of the current baseline, met few of the criteria that the Army felt were critical to successful utilization of the MAs to develop operational requirements. Approach 2 calls for modification of the baseline by defining an operational objectives level along with refining the Mission Areas. If implemented, this approach would require the definition of operational capabilities. Figure 5.1 shows that these do not currently exist with the box marked "No" for "Assess operational capability." Approach 2 does, however, provide a better ability to explain the operational demands the Army needs to

RAND *MR1133-5.1*

		Approach 1 Refine baseline	Approach 2 Modify baseline	Approach 3 Create hybrid
Criteria for Mission Areas/operational objectives	Explain Army	No	Yes	Yes
	Assess operational capability	No	No	Need all opn'l objectives
Criteria for tasks, capabilities	Complete	No	?	?
	Not redundant	No	Yes	Yes
	Operable and measureable	No / No	Yes / No	Yes / Not now
	Few in number	122 / 789	74 / 467	52 / 374

Figure 5.1—Assessment of Three Approaches

provide. Approach 3, the hybrid, is similar in its ability to identify operational capabilities and explain the Army.

The second level of assessment, shown in the lower half of the figure, summarizes our assessment of criteria for tasks and operational capabilities. The criteria—complete, not redundant, operable and measurable, and few in number—are shown second from the left. The assessment for each option is shown in the boxes. The qualitative evaluation reveals that completeness of the ability to capture the totality of tasks and capabilities are still not known for Approaches 2 and 3, given that the work has not been done. Approaches 2 and 3, however, solve the redundancy problem identified in Approach 1. Approaches 2 and 3 also provide some improvement in the areas of operable and measurable tasks and reducing the number of tasks.

We did not recommend a specific option: rather, we provided an array of alternative approaches and assessed their comparative advantages and disadvantages.

Performance Measures

To the extent that common metrics (scales) can be applied, they greatly simplify the performance standard and measurement process. Some form of readiness across MAs, objectives, and capabilities holds the most promise as a common indicator. For some MAs outlined above (e.g., Promote Regional Stability), a direct (but subjective) assessment of the outcome is possible, e.g., Bosnia is different thus far from Somalia. For other MAs (e.g., Fight and Win Major Theater Wars), direct assessment is seldom feasible. Instead, we use an appropriate proxy metric to assess this outcome. In most cases, readiness has become the metric for setting performance standards and measuring actual performance. The metric is the correct one, for reasons to be discussed below; but the currently used scale for readiness has well-documented flaws. For example, there is no universally acknowledged definition of readiness. The term is itself generic; it might have a different form when applied to different capabilities. For different missions or capabilities, the precise metric might be different (fits the particular capability), but it could be under the generic name of readiness. As a result, one could assert that readiness was up, down, or stable for each mission or capability and offer as evidence the particular metric germane to that mission or capability.

OSD views readiness as a major measure of merit of the competing priorities of modernization, ongoing mission responsibilities, and current readiness (DoD Annual Report, 1998). Readiness is a common language shared among the services, OSD, JS, CINCs, and Congress, but there is no standard definition. In part that is because standards for readiness are often externally defined in such documents as NMS, Chairman's Program Recommendation, DPG, and Joint Planning Document. OSD has also made readiness a GPRA Level 5 goal: Maintain highly ready joint forces to perform the full spectrum of military activities. The Army's own definition of force readiness fits well with the outlined concept for mission and capabilities analyses in that it ties resources of several types to missions.

> Force readiness is defined as the readiness of the Army within its established force structure, as measured by its ability to station, control, man, equip, replenish, modernize, and train its force in peacetime, while currently planning to call up, mobilize, prepare,

deploy, employ, and sustain them in war to accomplish assigned missions. (*How the Army Fights,* 1997–1998)

Thus, readiness is the primary objective of peacetime planning and programming. The goal is to maintain current readiness of capabilities for specific missions while ensuring future readiness of capabilities through investment. All of the functional enablers contribute to present and future readiness.

As noted above, readiness is hierarchical. One level of the hierarchy provides the inputs to the next level. For example, personnel readiness and materiel readiness feed unit readiness, which is itself a component of force readiness.[5] Readiness is measured both qualitatively and quantitatively. Quantitative information about inputs or intermediate outputs can form the basis, when combined with military judgment, for qualitative overall assessments about capabilities.

---

[5]John F. Schank, Margaret C. Harrell, Harry Thie, Monica M. Pinto, and Jerry M. Sollinger, *Relating Resources to Personnel Readiness: Use of Army Strength Management Models,* Santa Monica, CA: RAND, MR-790-OSD, 1997.

Chapter Six

CONCLUSIONS AND RECOMMENDATIONS FOR TAP 02–17

Mission Areas must focus on the external environment and cannot be static. MAs will change as national strategies and CINC missions change. The Title 10 institutional mission of the Army, the Army vision, the core competencies, imperatives, and the Army functions are more enduring. All of these must be shaped toward providing operational capabilities that aggregate into MAs.

Capabilities need to be operationally and doctrinally based. They result from the way the Army translates inputs and intermediate outputs into outcomes. Experience and knowledge in current and emerging Army doctrine and operations are needed to determine and assess capabilities.

MAs, operational objectives, and capability assessments must be conducted periodically. The best times to make these assessments are before and after key resource actions. When done before, they guide decisions about resource allocations. When done after, they portray results.

The current process is too complex and cumbersome. Very few, if any, organizations as large and differentiated as the Army can simultaneously allocate resources and determine outcomes of such decisions. The process must be iterative. Too many people are involved and those people involved tend to have a programmer's focus. Fewer people who are involved need to have an operational and outcome perspective (MA and capability "owners") in order to make operational assessments.

Finally, the responsibility for MAs, operational objectives, and capabilities should be divorced from the section within the DCSOPS that has the responsibility for producing TAP. The Strategy Directorate should be responsible for conducting MA analysis and inputting results to the resource constraints of TAP. MAs and operational capabilities relate more to the external demands identified by the CINCs, Joint Staff, and OSD than to internal Army functions of budgeting and programming.

DEVELOPMENT OF TAP 02–17

Since this work was done, attempts are being made to improve the fidelity of the operational tasks and performance measures. The Army is relying more on the Universal Joint Task Lists (UJTLs) to assist in the development of the tasks associated with individual MAs. The UJTLs, however, were developed by the JS for training exercises and often are not as operationally focused as might be required to identify operational requirements. In addition to the refinement of operational tasks, the Army is attempting to better define operational capabilities.

RECOMMENDATIONS FOR PERFORMANCE MEASURES IN TAP 02–17

The Army needs to use a hierarchical performance framework based on readiness for its mission and capability assessments. Such a framework is understood within the Army and allows integration of multiple inputs and intermediate outputs into a comprehensive whole. The concept is consistent with OSD, JS, the other services, and Congress. It enables the Army to logically articulate its planning priorities and programming decisions to external audiences and provides a common tableau for discussion.

Overall force readiness is the proper metric for assessing MAs. The metric needs to be elaborated for each MA. Legitimate performance standards need to be determined for each MA, and methods to qualitatively assess and/or quantitatively measure each MA need to be developed. Metrics, standards, and measures need to be cascaded to subordinate objectives and capabilities.

This can be an iterative process. Experts responsible for each capability, with doctrinal knowledge and experience about each capability, look horizontally across plans and programs to assess the suggested factors for each capability. Initially, we would expect the first assessments to be more qualitative than quantitative. Military judgment about a capability is married to objective data and information about the capability to produce the assessment. Over time, accepted quantitative metrics, measures, and standards for capabilities should emerge.

OVERALL RECOMMENDATIONS FOR TAP 02–17

Recommendations for TAP 02–17 are as follows:

- Reduce MAs to those that focus on operational missions and use them to explain the Army to external audiences and to inform internal Army resourcing to include POM deliberations.

- Realign and simplify the MA hierarchy—use operational objectives, operational capabilities, and operational tasks as subordinate levels of the Army MA hierarchy.

- Start MA assessments prior to publication of TAP using the current POM. Ensure that the MA assessments are part of an iterative process. Reduce the numbers of people and time to complete.

- Consider placing responsibility for the MAs and assessments in a different portion of the DCSOPS, e.g., DAMO-SS, to give it greater linkage to strategic planning and where doctrinal expertise and the joint and OSD perspective reside. Moving the task to another section of DCSOPS would allow for greater objectivity in developing and assessing MAs.

- Do not develop ASPG, TAP, and APGM concurrently, and insure that they are done in a progressive sequence over a longer period.

LINKAGES AMONG JV2010, AV2010, AND ASPG

The ASPG links to both the Joint and Army visions. Similar to JV2010 and AV2010, it focuses on attaining full-spectrum dominance. The guidance builds on AV2010 by defining full-spectrum dominance as the Army ability to fully support a wide span of missions—humanitarian assistance to MTWs, etc. The full-spectrum dominance theme is further expanded upon in identifying the Army's goal of first attaining mental agility by applying information technology across Army systems. Information technology will enable the Army to see the battlefield, knowing the location of its own elements, those of its allies, and those of the enemy. Once the Army has attained mental agility, it will then seek to attain physical agility by upgrading current systems, acquiring new ground systems that are faster and light, and experimenting with new organizational concepts.

CRITICAL DEMANDS PLACED UPON THE ARMY

The purpose of the ASPG is to identify the critical demands being placed on the Army. The demands emanate from two sources: (1) externally identified requirements and (2) internally generated initiatives. The demands can be either operational or institutional in nature. Most externally identified demands focus on operational requirements. External requirements can come from a number of places: OSD, the CINCs, Congress, etc. These requirements are specifically assigned to the Army, or the Army perceives that a requirement can only be met through Army-developed capabilities. For example, the Army provides many capabilities that support

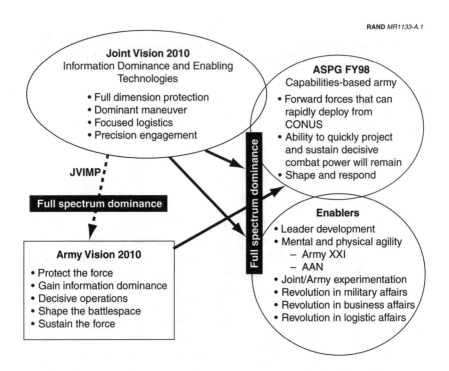

RAND *MR1133-A.1*

Figure A.1—Linkages Among JV2010, AV2010, ASPG

peacetime engagement missions. It also provides capabilities to prevent the spread of weapons of mass destruction (WMD). Both missions have been assigned to CINCs, who in turn look to the Army to provide unique sets of capabilities to meet the operational objectives and tasks associated with those missions.[1]

Institutional initiatives are often generated within the service. These requirements usually focus on those areas that ensure a service's ability to provide those capabilities that are unique to it. Institutional initiatives usually fall within such areas as man, train, equip,

---

[1]Some have argued that the division of Army requirements into two distinct areas creates two Armies. This is not so; rather, the distinction explicitly indicates that there are two important and essential elements to the creation of one Army, operational and institutional.

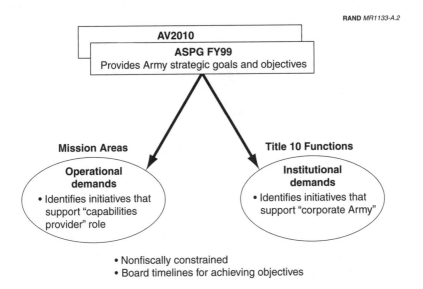

RAND *MR1133-A.2*

Figure A.2—ASPG Identifies Two Types of Demands

sustain, facilities, and organize; these are also referred to as the Title 10 functions because of their identification in U.S. legislation.

Title 10 functions cover a broad range of activities in that they ensure that the critical functions that underpin the Army core competency are addressed. For example, one Title 10 function is to ensure that the Army has a quality force. To carry out this function, the Army must train leaders and soldiers, and it must develop and field systems. These activities both indirectly and directly support its development and deployment of operational capabilities. Not all of the Army's force is directly involved in warfighting, for the force must also perform other activities. Thus, the development of a quality force means that the force must be capable of performing a wide array of warfighting and nonwarfighting activities. For example, the Army must ensure that its troops and their dependents receive proper housing, schooling, and medical care. This activity is critical to recruiting and sustaining a qualified force.

TRANSLATING GUIDANCE FROM THE ASPG TO TAP

The ASPG defines the key initiatives that define the Army's path to the future, that is, how the Army leadership wants changes to occur and the level of risk that it is willing to accept. The ASPG contains two types of initiatives: operational and institutional. We offer an example of each type and describe how they are implemented.

An operational initiative puts into motion the senior leadership's goal of providing efficient operational capability to the CINCs. The example used in this discussion revolves around the need to identify new operational concepts to support Army XXI. This need appears in the ASPG, and the initiative would then be coordinated by the DCSOPS, who has responsibility for the planning process. The initiative would be "operationalized" by tasking Training and Doctrine Command (TRADOC), which has the responsibility for the development of new concepts. TRADOC would begin to identify concepts through the battlefield functional Mission Area process, which would provide concepts, define battlefield and experimentation requirements, and finally coordinate with commands.

Once the leadership approves a concept, the Mission Area teams assess it in terms of its impact on the Army's ability to effectively provide future capabilities. If the concept is identified as critical in helping the Army to achieve a desired operational objective in the future, it is then identified as a resourcing requirement.

The institutional example falls within the Organize PEG and facilitates the broader goal of transitioning to Army XXI. The initiative is the redesign of the division that the Army leadership has determined is critical to the Army's achieving Army XXI. The initiative is institutional because it focuses on organization, not operations, although it will ultimately affect how Army operational capabilities are provided. For example, the division redesign also affects the institutional Army in such areas as manning, equipping, and training, which must be addressed by those PEGs for their respective resource requirements.

Although operational and institutional initiatives ultimately affect each other, it is important to note that they are handled independently. Independent treatment is important because it ensures that the Army leadership can make informed decisions about how to balance external and internal demands. New initiatives must compete for funding with ongoing ones, because the Army operates within a fixed budget. The Army programming structure has been reengineered to reinforce this independence, with the MA Teams (a new organization) dealing with the external demands of the CINCs, the six PEGs (a redesigned organization) addressing internal capabilities, and the Office of the AVCSA (a new organization) integrating the two.

MISSION AREA TABLES

Operational Mission Areas

Mission Area	DPG/QDR/FY98 SecDef Report	DoD Corp Goals	Pirnie	AV 2010	TAP	NMS 1997	FY97 SecDef Report	FY96 SecDef Report	FY95 SecDef Report
Promote regional stability	X				X	X			
Foster regional stability			X						
Reassurance				X					
Shape the international environment through engagement programs and activities		X							
Stability through overseas presence							X	X	X
Serve as role models for militaries in emerging democracies						X			

Mission Area	DPG/ QDR/ FY98 SecDef Report	DoD Corp Goals	Pirnie	AV 2010	TAP	NMS 1997	FY97 SecDef Report	FY96 SecDef Report	FY95 SecDef Report
Prevent or Reduce Conflicts and Threats	X					X			
Punitive intrusion				X					
Respond to transnational threats (terrorism, drug traffic, international organized crime)						X			
Counter regional threats involving WMD			X						
Counter the spread and use of WMD	X						X	X	X
Counter the proliferation of WMD						X			
Deter and counter state-sponsored and other terrorism	X		X						
Combating terrorism	X						X	X	X
Counter production and trafficking in illegal drugs	X		X						
Counter drug operations							X	X	X
Perform other missions					X				

Mission Area	DPG/ QDR/ FY98 SecDef Report	DoD Corp Goals	Pirnie	AV 2010	TAP	NMS 1997	FY97 SecDef Report	FY96 SecDef Report	FY95 SecDef Report
Deter Aggression and Coercion									
Deter aggression and coercion	X			X					
Deter aggression and coercion on a daily basis						X			
Deter aggression and coercion in crisis	X								
Deter and defeat aggression against U.S. allies, friends, and interests			X						

Mission Area	DPG/QDR/FY98 SecDef Report	DoD Corp Goals	Pirnie	AV 2010	TAP	NMS 1997	FY97 SecDef Report	FY96 SecDef Report	FY95 SecDef Report
Conduct SSCs									
SSC operations	X								
SSCs						X			
Smaller-scale combat operations							X	X	X
Shape the int'l environment and respond to the full spectrum of crises by providing appropriately sized and mobile forces		X							
Protect the lives of U.S. citizens in foreign locations			X						
Noncombatant evacuation operations	X							X	X
Conflict containment				X					
Conduct peacetime operations					X				
Peacekeeping and peace enforcement ops							X		
Peace operations (multilateral and unilateral)	X							X	X
Contingency operations	X						X	X	X
Humanitarian				X					
Provide humanitarian and disaster relief at home and abroad			X						
Humanitarian and refugee assistance							X	X	X
Humanitarian assistance	X					X			

Mission Area	DPG/ QDR/ FY98 SecDef Report	DoD Corp Goals	Pirnie	AV 2010	TAP	NMS 1997	FY97 SecDef Report	FY96 SecDef Report	FY95 SecDef Report
Fight and Win MTWs									
Fight and win MTWs	X			X					
Major theater warfare						X			
Deter & defeat aggression (2 MTWs)							X	X	X
Defending or liberating territory				X					
Win the land battle					X				
Decisive operations				X					

Mission Area	DPG/ QDR/ FY98 SecDef Report	DoD Corp Goals	Pirnie	AV 2010	TAP	NMS 1997	FY97 SecDef Report	FY96 SecDef Report	FY95 SecDef Report
Secure the Homeland*									
Homeland defense	X								
Deter and defeat attacks on the U.S.			X						
Core security				X					

*More recent additional sources also support this Mission Area.

Institutional Mission Areas

Supporting Institutional Function	DPG/QDR/FY98 SecDef Report	DoD Corp Goals	Pirnie	AV 2010	TAP	NMS 1997	FY97 SecDef Report	FY96 SecDef Report	FY95 SecDef Report
Exploit Conceptual Innovations and Modernize Forces Accordingly									
Exploit the RMA	X								
Exploit the RBA	X								
Foster innovation in new operational concepts, capabilities, technologies, and structures						X			
Pursue a focused modernization effort	X								
Modernize our forces						X			
Fundamentally reengineer the DoD & achieve a 21st century infrastructure by reducing costs while maintaining required military capabilities across all DoD mission areas		X							
Insure/hedge against unlikely, but significant, future threats	X								
Adequately prepare to respond more effectively to unlikely, but significant, future threats						X			
Prepare now for an uncertain future by pursuing a focused modernization effort that maintains U.S. qualitative superiority in key warfighting capabilities		X							
Prepare now for an uncertain future by exploiting the RMA to transform U.S. forces for the future		X							

Supporting Institutional Function	DPG/QDR/FY98 SecDef Report	DoD Corp Goals	Pirnie	AV 2010	TAP	NMS 1997	FY97 SecDef Report	FY96 SecDef Report	FY95 SecDef Report
Prepare forces and provide capabilities									
Improve conduct of joint/combined operations	X								
Gain information dominance				X					
Achieve dominant battlefield awareness & information dominance					X				
Project the force	X			X					
Sustain the force				X	X				
Deploy the force					X				
Generate the force					X				
Protect the force				X	X				
Leverage				X	X				

Outcomes	DPG/QDR/FY98 SecDef Report	DoD Corp Goals	Pirnie	AV 2010	TAP	NMS 1997	FY97 SecDef Report	FY96 SecDef Report	FY95 SecDef Report
Preparedness/readiness									
Maintain highly ready joint forces to perform the full spectrum of military activities		X							
Maintain force readiness					X				

STRAWMAN OPERATIONAL OBJECTIVES

Protect lives of U.S. citizens abroad

Participate in noncoercive peace operations (peacekeeping)

Participate in coercive peace operations (peace enforcement)

Provide humanitarian and disaster relief at home and abroad

Conduct low-intensity conflict operations

Exercise command and control

Provide intelligence, surveillance, and reconnaissance

Dominate opposing operations and operate at will

Destroy opposing stocks and infrastructure

Sustain forces

Manage casualties

Conduct posthostility operations

Ensure the survivability of U.S. nuclear weapons and their control

Defend the United States against opposing attacks using WMD

Protect infrastructure

Defend against information operations

Counter terrorism in continental United States (CONUS)

Counter drug trafficking in CONUS

Provide military support to civilian authorities

Conduct consequence operations

Structure the force

Educate and train the force

Equip forces

Mobilize

Deploy

Support forces

Identify equipment, modernization, and acquisition

Maintain information management capabilities

Develop operational concepts and doctrine

Use models and simulations to support organizing, training, equipping, projecting, and sustaining

Conduct research and development

Conduct testing and experimentation

STRAWMAN OPERATIONAL CAPABILITIES

Conduct show of force and other demonstrations

Conduct inspections

Interdict shipments

Rescue U.S. citizens held hostage

Conduct permissive and nonpermissive noncombatant evacuation operations

Report and resolve violations of agreements

Interpose force

Assist in maintaining civil order

Help to create (replace) or repair damaged infrastructure

Control movement within/across borders

Establish and protect safe areas for civilians

Enforce cease-fire, disengagement, and arms limitations

Suppress and destroy forces of recalcitrant parties

Provide humanitarian and disaster relief at home and abroad

Help friendly regimes combat insurgency support insurrection against hostile regimes

Provide assistance in civil wars in foreign countries

Conduct raids

Destroy terrorism bases and infrastructure

Locate, suppress, and destroy opposing WMD

Conduct punitive operations

Exercise command and control

Collect information

Process information

Disseminate information

Conduct opposed assaults

Repel opposing attacks

Maneuver friendly forces

Destroy opposing forces

Evict opposing forces

Maintain rear area security

Degrade opposing stocks of war-related products

Degrade opposing output of basic industrial goods

Disrupt opposing communications

Disrupt opposing power generation

Disrupt opposing transportation

Obtain host nation support

Provide ammunition and munitions

Provide petroleum products, rations, and other expendables

Replace weapons and equipment

Provide replacement personnel

Establish theater-level maintenance

Construct, repair, and maintain infrastructure

Conduct health service support

Manage casualties

Conduct posthostility operations

Ensure the survivability of U.S. nuclear weapons and their control

Defend the United States against opposing attacks using WMD

Protect infrastructure

Defend against information operations and counter terrorism in CONUS

Counter drug trafficking in CONUS

Provide military support to civilian authorities

Conduct consequence management operations

Develop structure

Authorize units and organizations

Educate and train individuals

Conduct unit training

Ensure interoperability

Conduct exercises

Provide and maintain weapons and materiel

Provide munitions

Prepare for mobilization

Prepare units and individuals for deployment

Mobilize CONUS sustaining base

Deploy

Manage human resources

Provide quality-of-life support

Establish maintenance support

Establish medical and health service support

Provide necessary base operations/infrastructure

Provide POL, rations, and expendables

Identify equipment, modernization and acquisition

Maintain information management capabilities

Develop operational concepts and doctrine

Use models and simulations to support organizing, training, equipping, projecting, and sustaining

Conduct research and development

Conduct testing and experimentation

HIERARCHY OF MISSION AREAS, OBJECTIVES, CAPABILITIES

1. Promote Regional Stability

Maintain Military-to-Military Contacts

Provide Assistance

Training assistance
Materiel assistance

Participate in Exercises

2. Deter or Reduce Conflicts or Threats

Maintain Presence

Station forces OCONUS
Deploy forces periodically

Defend and Protect U.S. and Allied Forces

WMD protection
Conventional protection
Terrorism protection

Conduct Show of Force and Other Demonstrations

Prevent Proliferation of WMD and Conventional Weapons

Conduct inspections
Interdict shipments

3. Conduct Smaller-Scale Contingency Operations

Protect Lives of U.S. Citizens Abroad

Rescue U.S. citizens held hostage

Conduct permissive and nonpermissive noncombatant evacuation operations

Participate in Noncoercive Peace Operations (Peacekeeping)

Report and resolve violations of agreements

Interpose force

Assist in maintaining civil order

Help to create (replace) or repair damaged infrastructure

Participate in Coercive Peace Operations (Peace Enforcement)

Control movement within and across borders

Establish and protect safe areas for civilians

Enforce cease-fire, disengagement, and arms limitations

Suppress and destroy forces of recalcitrant parties

Provide Humanitarian and Disaster Relief at Home and Abroad

Conduct Low-Intensity Conflict Operations

Help friendly regimes combat insurgency

Support insurrection against hostile regimes

Provide assistance in civil wars in foreign countries

Conduct raids

Destroy terrorism bases and infrastructure

Locate, suppress, and destroy opposing WMD

Conduct punitive operations

4. Fight and Win Major Theater Wars

Exercise Command and Control

Provide Intelligence, Surveillance, and Reconnaissance

Collect information

Process information

Disseminate information

Dominate Opposing Operations/Operate at Will

Oppose assaults

Repel opposing attacks

Maneuver friendly forces

Destroy opposing forces

Evict opposing forces

Maintain rear area security

Destroy Opposing Stocks and Infrastructure

Degrade opposing stocks of war-related products

Degrade opposing output of basic industrial goods

Disrupt opposing communications

Disrupt opposing power generation

Disrupt opposing transportation

Sustain Forces

Obtain host nation support

Provide ammunition and munitions

Provide petroleum products, rations, and other expendables

Provide replacement weapons and equipment

Provide replacement personnel

Establish theater-level maintenance

Construct, repair, or maintain required infrastructure

Conduct health service support

Manage Casualties

Conduct Post-Hostility Operations

5. Secure the Homeland

Ensure the survivability of U.S. Nuclear Weapons and their Control

Defend the U.S. against attacks using WMD

Protect Infrastructure

Defend Against Information Operations

Counter Terrorism in CONUS

Counter Drug Trafficking in CONUS

Provide Military Support to Civilian Authorities

Conduct Consequence Management Operations

6. Prepare Forces and Provide Capabilities

Structure the Force

> Develop structure
> Authorize units and organizations

Educate and Train the Force

> Educate and train individuals
> Conduct unit training
> Ensure interoperability
> Conduct exercises

Equip Forces

> Provide and maintain weapons and materiel
> Provide munitions

Mobilize

> Prepare for mobilization
> Prepare units and individuals for deployment
> Mobilize CONUS sustaining base

Deploy

Support Forces

    Manage human resources

    Provide quality-of-life support

    Establish maintenance support

    Establish medical and health service support

    Provide necessary Base Operations/infrastructure

    Provide POL, rations, and other expendables

7. Exploit Concept Innovation and Modernize Forces Accordingly

Identify Modernization and Acquisition Issues

Maintain Information Management Capabilities

Develop Operational Concepts and Doctrine

Use Models and Simulations

Conduct Research and Development

Conduct Testing and Experimentation

BIBLIOGRAPHY

Keeney, Ralph L., "Structuring Objectives for Problems of Public Interest," Operations Research, Vol. 36, No. 3, May-June 1988.

Kent, Glenn, *Concepts of Operations: A More Coherent Framework for Defense Planning*, Santa Monica, CA: RAND, N-2026-AF, 1983.

Kirkwood, Kraig W., *Strategic Decision Making*, Belmont, CA: Duxbury Press, 1997.

Lewis, Leslie, Bruce Pirnie, William Williams, and John Schrader, *Defining a Common Planning Framework for the Air Force*, Santa Monica, CA: RAND, MR-1006-AF, 1999.

Lewis, Leslie, and C. Robert Roll, *Strategy-to-Tasks: A Methodology for Resource Allocation and Management*, Santa Monica, CA: RAND, P-7839, 1993.

Lewis, Leslie, and Roger Brown, "Army POM, FY 00–05 Assessment," briefing to the Army Program Analysis and Evaluation, RAND, August 1998.

Lewis, Leslie, Roger Brown, and John Schrader, *Improving the Army's Resource Decisionmaking*, Santa Monica, CA: RAND, DB-294-A, 2000.

Lewis, Leslie, Roger Brown, and John Schrader, *Improving Army PPBES: The Programming Phase*, Santa Monica, CA: RAND, MR-934-A, 1999.

Office of the Secretary of Defense, *Annual Report to the President and Congress,* Washington, D.C.: U.S. Department of Defense, 1998.

Pirnie, Bruce, *An Objectives-Based Approach to Military Campaign Analysis,* Santa Monica, CA: RAND, MR-656-JS, 1996.

Schank, John, Margaret Harrell, Harry Thie, Monica Pinto, and Jerry Sollinger, *Relating Resources to Personnel Readiness: Use of Army Strength Management Models,* Santa Monica, CA: RAND, MR-790-OSD, 1997.

Shalikashvili, John, *Joint Vision 2010,* Washington, D.C., 1996.

U.S. Department of the Army, *Army Strategic Planning Guidance '99 Draft,* 1998.

U.S. Department of the Army, *How the Army Runs: A Senior Leader Reference Handbook, 1997–1998,* Carlisle Barracks, PA: 1997.

U.S. Department of the Army, *The Army Plan,* 1998.

U.S. Department of Defense, *Joint Vision Implementation Master Plan,* Washington, D.C., 1997.

U.S. Department of Defense, *Universal-Joint Task List,* Washington, D.C., 1995.

U.S. General Accounting Office, *Executive Guide: Effectively Implementing the GPRA,* Washington, D.C.: GAO-GGD—96-118, 1996.